William Shakespeare

U0108477

The Tempest

暴風雨

商務印書館

This Chinese edition of *The Tempest*
has been published with the written permission of
Black Cat Publishing.

The copyright of this Chinese edition is owned by
The Commercial Press (H.K.) Ltd.

Name of Book: The Tempest
Author: William Shakespeare
Adapted by: Victoria Heward
Editors: Claudia Fiocco, Rebecca Raynes
Design: Nadia Maestri
Layout: Simona Corniola
Illustrations: Alfredo Belli
Edition: ©2002 Black Cat Publishing
          an imprint of Cideb Editrice, Genoa, Canterbury

系 列 名：Black Cat 優質英語・漫畫莎劇
書　　名：暴風雨
顧　　問：Angeli Lau
責任編輯：黃淑嫻
封面設計：張　毅
出　　版：商務印書館（香港）有限公司
　　　　　香港筲箕灣耀興道3號東滙廣場8樓
　　　　　http://www.commercialpress.com.hk
發　　行：香港聯合書刊物流有限公司
　　　　　香港新界大埔汀麗路36號中華商務印刷大廈3字樓
印　　刷：中華商務彩色印刷有限公司
　　　　　香港新界大埔汀麗路36號中華商務印刷大廈
版　　次：2008年7月第2次印刷
　　　　　© 商務印書館（香港）有限公司
　　　　　ISBN 13-978 962 07 1693 5
　　　　　ISBN 10-962 07 1693 0
　　　　　Printed in Hong Kong

版權所有　不得翻印

出版説明

    莎劇，魅力無窮，自十六世紀以來，就是讀不盡的大書，演不完的戲劇，演出形式多種多樣，從話劇到歌劇，再到影視，真可以説，莎劇充滿了世界的每一個角落。莎劇中充滿智慧的名句，更成為人生的格言，歷久彌新。然而，莎翁原著畢竟是古英文，對一般讀者而言，往往比較艱深，難以領略其精妙。本館因此發愿，以創新的形式推動莎翁名劇普及，為新一代朋友提供走近大師的方便之門，此次即特別推出這套以莎士比亞名劇為主題的漫畫讀本。

    《漫畫莎劇》以漫畫的表現手法重現莎劇精髓，活靈活現的場景配上簡單精煉的人物對白，讓讀者目擊劇情的曲折發展，直面人物的內心世界，置身於莎士比亞豐富的戲劇天地。淺白的現代英語，生動的彩色圖像，使初級程度的讀者得以欣賞這位大文豪的不朽風采。

    此套書設計功能獨到，學習莎劇“動靜皆宜”。讀者既可根據個人愛好閱讀欣賞，增長有關戲劇的各種知識，掌握戲劇的基本元素，又可使用本系列作為排演指南及腳本，透過書內設置的豐富多樣的小組活動和練習，實際參與演戲或幕後工作，在實踐中體驗莎劇的精妙。

    本館一向倡導優質閱讀，近年來連續出版以“Q”為標識的“Quality English Learning 優質英語學習”系列，譬如《讀名著學英語》、《Black Cat 優質英語階梯閱讀》等，深受讀者歡迎。莎翁名劇是經典中的經典，本館期望透過嶄新的漫畫手法，推動經典藝術的普及，讓莎劇的精神薪火相傳。

商務印書館（香港）有限公司
編輯部

使用說明

① 如何使用本系列？

Step One:

本系列提供莎士比亞的生平簡介，部分附設劇情提要，讀者宜先略讀這兩部分，以便對該劇有初步了解。

Step Two:

從"劇中人物表"掌握各主要人物的身份、地位及彼此之間的親屬關係。在閱讀過程中若讀者遇到個別生詞，即可翻到書後的"實用詞彙表"查找。若不明白個別戲劇用語，亦可翻到書後的"戲劇詞彙表"查閱。

Step Three:

每本書都按劇情分成不同場景（Acts 或 Parts），每個場景後跟多樣化的練習，一部分有助理解劇情，一部分教莎翁時代的英語，還有一部分教演戲技巧，包括揣摩角色、自己動手製作道具及戲服、替演員化妝、製造音效以及佈置舞台等等。

② 本系列的練習有何功能？

本系列的練習分三大類，第一類針對內容，讓讀者自測能否充分理解劇情的發展和人物的關係；第二類針對學習英語，特別集中教英語語法及詞彙，例如讓讀者判斷哪些形容詞（adjectives）能夠與劇中人物的性格配對，又備有反義詞（opposites）、詞語後綴（suffixes）、押韻（rhyme）等練習；第三類是針對訓練戲劇技巧的活動，例如引導讀者以身體不同的動作模仿海浪，又教讀者學習用不同的聲線語調扮演劇中人物等，均能幫助讀者從實際的參與學會台前幕後的各項技巧。

③ 本系列能提供的其他幫助？

每本書內配有專業外籍演員錄製的CD，可供讀者摹仿朗讀及完成聽力練習。此外，本系列提供莎士比亞時代的歷史、文化及社會背景，例如列舉文藝復興時代的社會潮流及著名的代表人物，說明劇中主角在歷史上真有其人及介紹故事發生地的地理環境，又提供莎翁劇院（The Globe）的相關資料。另外，本系列具備一些趣味小知識，如當時的女角均由男人反串飾演，蘇格蘭的姓氏多以Mac開頭等，以幫助讀者理解內容及提高閱讀趣味。

CONTENTS

William

SHAKESPEARE

Name:
William Shakespeare

Born:
In Stratford-Upon-Avon, England

When:
23 April 1564

Most famous plays:
Hamlet
Macbeth
Romeo and Juliet

Died:
23 April 1616

Project

Shakespeare is the most famous English writer.
Who is the most famous writer in your country?

Try to find out more information about Shakespeare.

Find out about: a. his life
                                b. his plays
                                c. English history during his life.

Use one of the following search engines:
http://www.altavista.com
http://www.searchalot.com
http://www.yahoo.com

CAST
劇中人物

ALONSO　'I am the King of Naples.'

那不勒斯國王阿隆索

FERDINAND　'I am the King's son.'

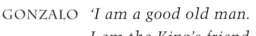

SEBASTIAN　'And I am the King's brother.'

那不勒斯王子
斐迪南

ANTONIO　'I am the Duke of Milan and
I am Prospero's brother.
I'm not a good man.'

阿隆索之弟西巴斯辛

GONZALO　'I am a good old man.
I am the King's friend.'

普洛斯彼羅之弟安
東尼奧，篡位者

正直的老臣貢薩羅

PROSPERO　'I am the real
Duke of Milan and
I'm a magician.'

舊米蘭公爵普洛斯彼羅

MIRANDA　'I'm Prospero's daughter.'

普洛斯彼羅的女兒米蘭達

CALIBAN　'I'm a monster.' [1]

ARIEL　'And I'm a magic spirit.' [2]　半獸半人的怪物卡利班

小精靈愛麗兒

STEPHANO　'I'm a butler. [3] I work
on the King's ship.'

TRINCULO　'I'm a jester [4] and I work
on the King's ship too.'

船上的僕役長
斯丹法諾

船上的小丑特林鳩羅

UNDERSTANDING THE TEXT

KET

1 For questions 1-5, tick ✔ A, B or C.

1. Prospero wants

     A ☐ to live on an island.

     B ☐ to see a monster.

     C ☐ to do some magic.

2. Alonso lives in

     A ☐ England.

     B ☐ Italy.

     C ☐ France.

3. On the island there is

     A ☐ a monster and a spirit.

     B ☐ a big castle.

     C ☐ a pirate.

4. Caliban is

     A ☐ not a slave.

     B ☐ Prospero's slave.

     C ☐ Antonio's brother.

5. Miranda and Prospero live on the island for

     A ☐ five years.

     B ☐ two years.

     C ☐ twelve years.

2 Answer these questions.

a. Where does Ariel live?

b. Who lives in a castle?

c. Who has a bad brother?

d. What does Prospero do every day?

e. Who is the Duke of Milan?

3 Complete the weather chart by adding the correct ending under each example.

Adjective	V + ing
su.*nny*.......................	rain..*ing*..........................
fog.............................	snow ....
cloud.........................	hail ....
wind..........................	drizzle .... (light rain)
storm.........................	

---------- KET ----------

4 **Now listen to the conversation and fill in the chart.**
You will hear the conversation twice.

Today is a | 0. *beautiful*............... | day.

It is | 1. .... | but it is also | 2. .... | .

Ariel thinks tomorrow it will | 3. .... | .

The weather forecast⁷ says that tomorrow it will | 4. .... |
on the mountains.

Near the | 5. .... | it will be sunny.

In the afternoon it will be | 6. .... | .

The next day, stay at home because it will be very | 7. .... | .

A tempest is a | 8. .... | .

5 **Answer these questions.**

a. What's your favourite type of weather? I like .... weather.

b. What do you like to do in the summer? I like ....ing.

c. What do you like to do in the winter? I like ....ing.

d. What's the weather like today? It is ....

e. When is your birthday? What's the weather usually like on your birthday? ....

14

UNDERSTANDING THE TEXT

1 Listen to the conversation. For questions 1-8, tick ☑ A, B or C. You will hear the conversation twice.

1. It is
- ☐ **A** morning.
- ☐ **B** afternoon.
- ☐ **C** evening.

2. What time is it?
- ☐ **A** 6.30.
- ☐ **B** 7.00.
- ☐ **C** 7.30.

3. Today is not a good day because
- ☐ **A** Antonio is sleeping.
- ☐ **B** a storm is coming.
- ☐ **C** the King is angry.

4. The sky is
- ☐ **A** blue.
- ☐ **B** big.
- ☐ **C** black.

5. The sea is
- ☐ **A** hungry.
- ☐ **B** black.
- ☐ **C** angry.

6. Ferdinand is
- ☐ **A** a prince.
- ☐ **B** an old man.
- ☐ **C** a slave.

7. Gonzalo is
- ☐ **A** in bed.
- ☐ **B** working on the ship.
- ☐ **C** reading a letter.

8. Antonio and Alonso shout 'Help!' because
- ☐ **A** they see a monster.
- ☐ **B** the storm arrives.
- ☐ **C** they can't swim.

1 Match these verbs to one of the words opposite.

1. ☐ pull   a. pirates
2. ☐ salute   b. the television
3. ☐ watch   c. the ship
4. ☐ look for   d. the dinner
5. ☐ clean   e. the rope
6. ☐ cook   f. the captain

2 Which of the things above do you do on a ship?

3 Imagine you are on a big boat.
Mime [11] all the jobs in question one. The sea is calm.

IT'S VERY WINDY TODAY.

4 Be careful.
Your boat starts to move
in the wind.
It's difficult to work.
Now mime the jobs in the wind.

LOOK AT THE BIG WAVES.

5 Be careful.
The waves are very big now.
It's difficult to walk.
It's difficult to stand up.
Now mime the jobs in the storm.

6 But remember that the storm is a magic storm. Prospero controls everything.

THERE'S A BIG STORM TODAY.

a. One of the group members is Prospero and the others are the wind. The wind follows Prospero's directions: it pushes and then it pulls, it goes fast and then it goes slow.

b. One of the group members is Prospero and the others are the waves. The waves are calm at first and then they grow big. They go up and then they go down.

7 Listen to the music. Close your eyes and imagine a small boat in the tempest.

8 Divide the members in three groups.

- Group 1 is the boat.
- Group 2 is the wind.
- Group 3 are the waves.

9 Mime the tempest together with the music.

23

ACTING 透過唱歌拍掌認識節奏韻律

 1 **Listen and sing Ariel's song.**

Time to wake up
Time to wake up
Ferdinand
Come with me.
Time to do some magic
Time to do some magic
One, two, three
One, two, three

2 **When you see an 'x', clap your hands. When you see an 'o', click your fingers.**

Time to wake up
  X o X   o

Time to wake up
  X o X   o

Ferdinand
  X o X   o

Come with me.
  X   o X o

Time to do some magic
  X   o   X o

Time to do some magic
  X   o   X o

One, two, three
  X   o X   o

One, two, three
  X   o X   o

3 Now try this rhyme.
Half of the group members ask the question and the other half answer.

What are you doing Ferdinand?

I'm sleeping

I'm sleeping.

What are you doing Miranda?

I'm reading

I'm reading.

What are you doing Prospero?

I'm eating

I'm eating.

What are you doing Stephano?

I'm cleaning

I'm cleaning.

4 Here's another rhyme.
Choose the best word and then say the rhyme together.

> in   me   is   on   am   not   are   am

a. Where .... I?

b. Where .... I?

c. Where .... my friends?

d. And where .... my boat?

e. You're .... an island.

f. Your boat's .... the sea.

g. Your friends are .... here.

h. So, please come with ....

Part Four

UNDERSTANDING THE TEXT

1 Are these sentences 'right' (A) or 'wrong' (B)? If we don't know, choose 'Doesn't say' (C).

1. It is night time when they wake up.

    A ☐ Right   **B** ☐ Wrong   **C** ☐ Doesn't say

2. Sebastian has an idea to kill his brother.

    A ☐ Right   **B** ☐ Wrong   **C** ☐ Doesn't say

3. Gonzalo hears their conversation.

    A ☐ Right   **B** ☐ Wrong   **C** ☐ Doesn't say

4. Gonzalo is the King's friend.

    A ☐ Right   **B** ☐ Wrong   **C** ☐ Doesn't say

5. Ariel wants to help the King and Gonzalo.

    A ☐ Right   **B** ☐ Wrong   **C** ☐ Doesn't say

6. Prospero does some magic.

    A ☐ Right   **B** ☐ Wrong   **C** ☐ Doesn't say

7. Antonio is the new King of Naples.

    A ☐ Right   **B** ☐ Wrong   **C** ☐ Doesn't say

8. There are wild animals on the island.

    A ☐ Right   **B** ☐ Wrong   **C** ☐ Doesn't say

9. Antonio wants to protect King Alonso.

    A ☐ Right   **B** ☐ Wrong   **C** ☐ Doesn't say

10. Alonso wants to know where Ferdinand is.

    A ☐ Right   **B** ☐ Wrong   **C** ☐ Doesn't say

2   **What does Ariel say to Prospero?**
Complete the conversation and put the correct letter in the blank.

Prospero : Hello Ariel.
Ariel : ¹ .....
Prospero : What's the problem, Ariel?
Ariel : ² .....
Prospero : Oh! What's the matter?
Ariel : ³ .....
Prospero : A plan? What do they
     want to do?

Ariel : ⁴ .....
Prospero : Do they want to kill
     Gonzalo too?
Ariel : ⁵ .....
Prospero : Are they dangerous?
Ariel : ⁶ .....
Prospero : We must do something to stop them before it is too late.
Ariel : ⁷ .....
Prospero : Right, Ariel. Let's go!

A   Good idea, Sir.

B   They want to kill the King.

C   Yes, they do.

D   I have some terrible news for you.

E   Hello, Sir!

F   Yes, they are. They have big swords.

G   Sebastian and Antonio have a plan.

ACTING   上演刺殺國王事敗的一幕

3   **In Part Four Antonio and Sebastian want to kill the King and Gonzalo. Luckily they wake up! In a group, act Part 4 together. You need 5 people:**

Alonso   Antonio   Sebastian   Gonzalo   Ariel

Now try again and this time act a different part.

Part Five

OH, POOR ME!
POOR CALIBAN!
EVERY DAY I MUST
WORK FOR PROSPERO.
I WANT TO BE FREE,
BUT I AM HIS SLAVE.

WHERE ARE
THE OTHERS?
WHERE IS THE KING
AND WHERE IS
STEPHANO?

OH NO!
I HEAR A VOICE.
IT IS PROBABLY [17]
A BAD SPIRIT
FROM PROSPERO.
I MUST HIDE. [18]

1 Now you act the scene. In groups, act Part 5 together. You need 4 people:

> Caliban   Trinculo   Stephano   Ariel

CALIBAN: Oh, poor me!
Poor Caliban!
Every day
I must work
for Prospero.
I want to be free,
but I am his slave.

TRINCULO: Where are the others?
Where is the King
and where is Stephano?

CALIBAN: Oh no!
I hear a voice.
It is probably
a bad spirit
from Prospero.
I must hide.

TRINCULO: What a
beautiful island!
And what is this?
It has arms
and legs,
but no head.
It is a monster!
Oh no!
Another storm.
I must hide.

STEPHANO: Where are the others?
Where is the King
and where is Trinculo?
And what is this?
It has four arms
and four legs
but no head.
It is a monster!
Hello, monster.
Hello!

Now try again and this time act a different part.

CALIBAN: Help!
Prospero's bad spirits
want to torment me.

TRINCULO: Stephano!
Hello!
It's me,
Trinculo.

STEPHANO: And who is he?

CALIBAN: I am poor Caliban.
I am Prospero's slave,
but I want to be free.
I want to kill Prospero.

STEPHANO: That's terrible!
You must do something.
We can help you.

TRINCULO: No, Stephano,
it's not a good idea.

STEPHANO: Yes, we must kill
Prospero
and take control of the island.

CALIBAN: Thank you, thank you.

ARIEL: I must tell Prospero
of this terrible plan.

UNDERSTANDING THE TEXT

1   **What happens first?**
    Put these sentences in the correct order.

a. It starts to rain. ☐

b. They all decide to kill Prospero. ☐

c. Stephano sees a monster. ☐

d. Trinculo hides under Caliban's cloak. ☐

e. Caliban hides under his cloak. ☐

KET

2   **Read this postcard and fill in the blanks.**
    Write one word for each space.

Dear Mother,

How **0** .are..... you? I'm fine. **1** ....'s Saturday and it's four o'clock.
I am **2** .... a beautiful island. I am **3** .... a forest and I **4** ....
see lots of beautiful plants and **5** .... very strange animal. The animal
6 .... four arms and four legs, **7** .... no head.

I **8** .... go now, Mother because it's starting **9** .... rain.

Lots **10** .... love
your son,
Stephano

3   **Fill in the information about the postcard.**

a. Place : an island ....

b. Day : ....

c. Time : ....

d. He can see : ....

e. Number of legs : ....

f. Number of fingers : ....

g. Weather : ....

41

4 Today Caliban is ...
   Match the word to the picture.

   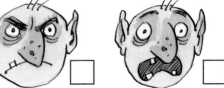

   a. afraid
   b. sad
   c. happy
   d. angry
   e. tired

5 Match the following sentences to the faces.

   a. ☐ Oh, poor me!
   b. ☐ Thank you, thank you!
   c. ☐ Every day, I must work.
   d. ☐ Oh no! I hear a voice. I must hide.
   e. ☐ I want to be free. I want to kill Prospero.

6 Label the pictures with the words in the box.

   hair   eyes   dress   jacket   shirt   hat

....

....

....

....

....

....

7   **Match these adjectives to their opposites.**

sad •   • ugly

old •   • stupid

pretty •   • happy

fat •   • poor

small •   • thin

rich •   • young

good •   • bad

intelligent •   • big

8   **Write sentences to describe these characters.**

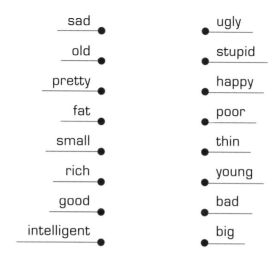

Miranda : Miranda is young and very pretty.

She has long hair and wears a blue dress.

Alonso : ....

....

Caliban : ....

....

Trinculo : ....

....

[The sound of music and Ariel vanishes.]

UNDERSTANDING THE TEXT

1 For questions 1-5, tick ✓ A, B or C.

1. Gonzalo wants
   - A ☐ to stop walking.
   - B ☐ to sleep.
   - C ☐ to kill Sebastian.

2. The food is
   - A ☐ not very good.
   - B ☐ magic food.
   - C ☐ a present from Miranda.

3. Ariel says,
   - A ☐ 'Stop eating!'
   - B ☐ 'You are bad men.'
   - C ☐ 'I am very sorry.'

4. Sebastian and Gonzalo think that
   - A ☐ Ariel is a monster.
   - B ☐ Miranda is poor.
   - C ☐ it's a good idea to eat the food.

5. Antonio is
   - A ☐ very tired.
   - B ☐ very angry.
   - C ☐ very sorry.

2 For questions 1-6, tick ✓ A, B or C.
You will listen to the conversation twice.

1. It's time to
   - A ☐ sleep.
   - B ☐ read.
   - C ☐ eat.

2. Alonso's favourite food comes from
   - A ☐ England.
   - B ☐ America.
   - C ☐ Italy.

3. Gonzalo has problems with his
   - A ☐ feet.
   - B ☐ teeth.
   - C ☐ wife.

4. Sebastian likes
   - A ☐ fish, soup and ice-cream.
   - B ☐ chocolate, chips and soup.
   - C ☐ ice-cream, fish and chips.

5. Gonzalo thinks it's a good idea to eat fish because
   - A ☐ he likes fish.
   - B ☐ they are on an island.
   - C ☐ there is no ice-cream.

6. To make chips you need
   - A ☐ fish.
   - B ☐ tomatoes.
   - C ☐ potatoes.

ACTING 訓練模仿技巧和想像力

1   Look at your partner and imagine you are looking into a mirror.
    When you move, your partner must copy you.

2   Now one person is the leader and the others must follow.
    Copy all of the leader's movements.

3   Listen to some different types of music, (it's better if there are no
    words): classical music, slow music, sad music, happy music.

4   Close your eyes and imagine that the music is telling a story.
    What is the story?

5   Imagine the music in a film and you are the actors.
    In groups, invent the story and then mime it together.

UNDERSTANDING THE TEXT

1 **Put the words into the right order.**
   What happens first?

   a. ☐1 the time is now it organise
      to wedding.

   ....

   b. ☐ some do to magic time.

   ....

   c. ☐ miranda very I love much.

   ....

   d. ☐ I it it beautiful want is.

   ....

   e. ☐ think stop must I dancing.

   ....

   f. ☐ now make will magic I some dogs.

   ....

2 **Can you remember who says the above words?**

   a. .Prospero.........................   b. ....

   c. ....   d. ....

   e. ....   f. ....

3 **Correct these sentences.**

   a. Caliban loves Miranda.

   b. Prospero wants to organise a concert.

   c. Stefano and Trinculo like
      the dogs.

   d. Ariel has a plan.

   e. Caliban does some magic.

5   Work with a partner. You must both choose to be different
     characters from the story. For example, A is Caliban and B is
     Ferdinand. Complete this dialogue together.
     NB Sometimes you have to use your imagination and invent answers.
     There is not one correct answer but lots of different alternatives.

     A: Hello, What's your name?
     B: ....
     A: Oh, that's a nice name. My name's ....
     B: Where do you come from?
     A: ....What about you?
     B: ....
     A: Very nice.
     B: What are you doing here?
     A: .... What about you?
     B: I ....
     A: I must go now because

              ....
     B: OK. Bye!

6   Now read the dialogue together and try to speak with the voice of
     your character.

7   Try the dialogue again but this time invent completely new
     characters. Don't tell your partner who you are or where you come
     from, etc. If you are A, complete only the dialogue for A. If you are
     B, complete the dialogue only for B. When you are ready, read the
     dialogue with your partner and see the result. Is your conversation
     funny or serious?

8   Finally work with your partner and use the dialogue to invent a
     dramatic situation. Who are the two people? Why are they there?
     What are they doing? Continue the dialogues.

Part Eight

There is a lot of noise and the dogs chase Caliban, Stephano and Trinculo. They stop in the middle of the royal[27] party.

WRITE A PLAY

1   Label the pictures.

> mobile phone   castle   island   key   moon   book
> thunder and lightning   envelope   diamond necklace

2   In small groups, choose and draw 3 more objects.
    Write the words in English.

1.	5.	9. ....
2.	6.	10. ....
3.	7.	11. ....
4.	8.	12. ....

3 Now you have 12 objects.
   Which objects are the most interesting?

   a. In your group choose 6 objects.
   b. Invent a story with all of these objects.
      You can speak about your story in your
      own language.

4 Imagine your story is a film or a play and
   you are the actors.
   Write the dialogue for your story in English.

   Example:

   Tom: *Look Anna. There is a gold key under the table.
   Let's take it.*

   Anna: *Good idea. I think it opens the door to the castle.
   Why don't we try?*

   Tom: *OK, but we must be quick because...*

5 In your group try to act out your story.
   Can you remember the words without looking at your notes?

6 Make a list of

   a. Costumes (special clothes for the actors).
   b. Props (objects to help tell the story).
   c. Music or sound effects.

7 Can you find or make these things for the next session?

   ## YOUR SHOW

8 Practise your play with costumes,
   props and sound effects.
   Which play do you like best? Why?

GLOSSARY　實用詞彙表

1　monster：怪物。
2　spirit：精靈。
3　butler：僕役長。
4　jester：小丑。
5　slaves：奴隸。
6　tempest：暴風雨。
7　weather forecast：天氣預測。
8　Your Majesty：陛下。
9　intelligent：聰慧的。
10　shipwreck：海難。
11　mime：模仿。
12　excellent：最好的。
13　opportunity：機會。
14　terrible：可怕。
15　sword：

16　protect：保護。
17　probably：可能。
18　hide：躲藏。
19　torment：折磨。
20　fault：過失。
21　organise：籌備。
22　wedding：婚禮。
23　snaps：彈（手指）。
24　couple：一對情侶。
25　scare：驚嚇。
26　celebrate：慶祝。
27　royal：王室的。
28　master：主人。
29　hurray：好哇（表示高興的歡呼聲）。
30　spell：魔咒。

GLOSSARY OF DRAMATIC TERMS　戲劇詞彙表

Cast : the group of actors in a show. 劇中人物。

Costumes : the clothes that the actors wear to seem more realistic. 戲服。

Director : the person who makes the decisions and tells the actors how to speak and where to move. 導演。

Dress rehearsal : the last rehearsal before the opening night. Everybody is in costume, with lights, props, sound effects etc. 總綵排。

Play : a story told by actors in a theatre. 戲劇。

Prompt : the person who stays hidden during the performance but who helps the actors if they forget their words. 提詞。

Props : the objects used on stage by the actors. 道具。

Scenery : part of the stage specially painted or constructed to represent the place where the story takes place. 舞台佈景。

Script : the text which contains the words that an actor must say, and some stage directions. 劇本。

Sound effects : noises which help create atmosphere like music, weather or shouting. 音響效果。

Stage : the place where the actors act. 舞台。

EXIT TEST

1 **Choose one of these words to complete this summary of Part One to Part Four of the story. Use each word once only.**

and / make / storm / son / ship / bad / they / happy / punish / island / waves / of / meets / stops

Part One:

Prospero is the Duke **1**[_____] Milan. Antonio is his **2**[_____] brother. Prospero wants to do some magic.
He wants to **3**[_____] his brother. He makes a big **4**[_____].

Part Two:

The King and the Duke are on a **5**[_____] but the **6**[_____] are very big and there is a lot of wind.
Prospero is very **7**[_____] with his magic.

Part Three:

The King and the others are sleeping on the **8**[_____]. Ferdinand wakes up **9**[_____] follows Ariel's song.
He **10**[_____] Miranda and they fall in love.

Part Four:

Sebastian and Antonio **11**[_____] a plan to kill the King and Gonzalo when **12**[_____] are sleeping. Luckily Ariel arrives and **13**[_____] this terrible plan. The King wants to look for his **14**[_____], Ferdinand.

2   Now choose one of these words to complete this summary of Part
    Five to Part Eight of the summary. Use each word once only.

> bad / help / wedding / food / he / brother /
> island / dogs / monster / eat / see

Part Five:

Caliban is a **15**[＿＿＿＿＿＿]. He must work for Prospero every day and
16[＿＿＿＿＿＿] is not happy. He meets Trinculo and Stefano. When
they hear the story of Caliban, they decide to **17**[＿＿＿＿＿＿] him kill
Prospero.

Part Six:

Lots of magic **18**[＿＿＿＿＿＿] appears. Alonso, Gonzalo, Sebastian and
the King want to **19**[＿＿＿＿＿＿] the food because they are very
hungry. When they try to take it, everything disappears. Ariel arrives
and tells them that they are **20**[＿＿＿＿＿＿] men.

Part Seven:

Prospero starts to organise the **21**[＿＿＿＿＿＿] for Miranda and
Ferdinand and he also sends magic **22**[＿＿＿＿＿＿] to chase Caliban,
Trinculo and Stefano.

Part Eight:

Finally everybody meets, and Prospero forgives his **23**[＿＿＿＿＿＿] and
the others. The King is very happy to **24**[＿＿＿＿＿＿] his son and he is
happy about the wedding too. Prospero says that now Caliban and
Ariel can be masters of the **25**[＿＿＿＿＿＿] .

The Tempest

KEY TO THE ACTIVITIES

UNDERSTANDING THE TEXT

Page 13 - 1
1. C / **2.** B / **3.** A / **4.** Two answers are possible A or B, depending on the point of view (Miranda says he's not a slave and Prospero says he is). / **5.** C

Page 13 - 2
a. Ariel lives on the island.
b. Antonio lives in a castle (Also possible 'The King/ Alonso lives in a castle' as kings usually do).
c. Prospero has a bad brother.
d. Prospero reads his (magic) book every day./ Prospero does magic every day.
e. Antonio is the Duke of Milan. (Here it is also possible to answer 'Prospero is the Duke of Milan' as we know that Antonio has usurped his brother).

WEATHER

Page 14 - 3
Sunny, foggy, cloudy, windy, stormy, raining, snowing, hailing, drizzling

LISTENING

Page 14 - 4
1. sunny / **2.** windy / **3.** rain / **4.** snow / **5.** sea / **6.** cloudy / **7.** stormy / **8.** big storm

Page 14 - 5
Open answers.

UNDERSTANDING THE TEXT

Page 19 - 1
1. A / **2.** C / **3.** B / **4.** C / **5.** C / **6.** A / **7.** A / **8.** B

ACTING

Page 20 - 1
1. e / **2.** f / **3.** b / **4.** a / **5.** c / **6.** d
Page 20 - 2
You can do all of them but *watch the television* is unlikely
Page 20 - 21, 3 - 9
Open answers.
Page 26 - 27, 1 - 3
Open answers.
Page 27 - 4
a. am / **b.** am / **c.** are / **d.** is / **e.** on / **f.** in / **g.** not / **h.** me

UNDERSTANDING THE TEXT

Page 33 - 1

1. C / **2.** B (It's Antonio's idea) / **3.** B
(Gonzalo is sleeping) / **4.** A / **5.** A /
6. B (Ariel does some magic) /
7. B (Antonio is the Duke of Milan) /
8. C / **9.** B (Antonio wants to kill the
King.) / **10.** A

Page 34 - 2

1. E / **2.** D / **3.** G / **4.** B / **5.** C / **6.** F / **7.** A

Page 34 - 3

Open answers.

Page 40 - 1

Open answers.

UNDERSTANDING THE TEXT

Page 41 - 1

a. 2 / **b.** 5 / **c.** 4 / **d.** 3 / **e.** 1

Page 41 - 2

1. It / **2.** on / **3.** in / **4.** can / **5.** a /
6. has / **7.** but / **8.** must / **9.** to / **10.** of

Page 41 - 3

b. Saturday / **c.** four o'clock / **d.** a
strange animal/ an animal/ beautiful
plants / **e.** 4 / **f.** 20 (possibly 16 if
thumbs are not counted as fingers) /
g. rainy/ it's raining

Page 42 - 4

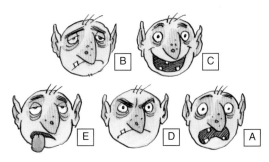

Page 42 - 5

a. sad / **b.** happy / **c.** tired / **d.** afraid /
e. angry

Page 42 - 6

Page 43 - 7

sad - happy, old - young, pretty - ugly,
fat - thin, small - big, rich - poor, good
- bad, intelligent - stupid

Page 43 - 8

Possible answers:

Alonso: Alonso is fat. He's got dark
hair and he always wears a cloak.

Caliban: Caliban is very ugly. He's
got a big nose and big ears.

Trinculo: Trinculo is always very
happy. He is wearing a hat.

UNDERSTANDING THE TEXT

Page 49 - 1

1. a / **2.** b / **3.** b / **4.** c / **5.** c

Page 49 - 2

1. c / **2.** c / **3.** b / **4.** c / **5.** b / **6.** c

Page 50, 1 - 5

Open answers.

UNDERSTANDING THE TEXT

Page 56 - 1

a. 1 - Now it is time to organise the
    wedding.

b. 4 - Time to do some magic.

c. 2 - I love Miranda very much.

d. 5 - It is beautiful. I want it.

e. 3 - Stop dancing. I must think.

f. 6 - Now I will make some magic dogs.

Page 56 - 2

b. Ariel / **c.** Ferdinand / **d.** Stephano / **e.** Prospero / **f.** Prospero

Page 56 - 3

a. Ferdinand loves Miranda.

b. Prospero wants to organise a wedding.

c. Stephano and Trinculo don't like the dogs.

d. Prospero has a plan.

e. Ariel does some magic./ Prospero does some magic.

Page 57 - 5

Possible answers:

A=Ferdinand / B=Caliban

A: Hello, What's your name?

B: My name's Caliban.

A: Oh, that's a nice name. My name's Ferdinand.

B: Where do you come from?

A: I come from Milan. What about you?

B: The island.

A: Very nice.

B: What are you doing here?

A: I am looking for my father. / I am lost. What about you?

B: I live here. I'm Prospero's slave. / I work here.

A: I must go now because I can hear some beautiful music. / I want to find my father.

B: OK. Bye!

Page 57, 6 - 8

Open answers.

Page 64 - 2

1. envelope
2. mobile phone
3. book
4. (open answer)
5. castle
6. diamond necklace
7. key
8. (open answer)
9. island
10. thunder and lightning
11. moon
12. (open answer)

Page 65, 3 - 8

Open answers.

TAPESCRIPT

TRACK 1

The Tempest by William Shakespeare
Adapted by Victoria Heward
(The words in italics are stage directions and must not be said.)
Alonso: I am the King of Naples. *(The King should sound rather pompous.)*
Ferdinand: I am the King's son. *(Ferdinand has a young voice. He's about 16 or 17.)*
Sebastian: And I am the King's brother. *(Sebastian is a bit of a villain. Antonio's sidekick.)*
Antonio: I am the Duke of Milan and I am Prospero's brother. I'm not a good man. *(Antonio is very dastardly. He is like a villain from a film who twirls his moustache and chuckles at his own wickedness.)*
Gonzalo: I am a good old man. I am the King's friend. *(Gonzalo inspires trust and affection.)*
Prospero: I am the real Duke of Milan and I'm a magician. *(Prospero is an old man who speaks with mystery*

and authority in his voice.)
Miranda: I'm Prospero's daughter.
(Miranda is the same age as Ferdinand.)
Caliban: I'm a monster.
Ariel: And I'm a magic spirit. *(Ariel is a female spirit.)*
Stephano: I'm a butler. I work on the King's ship.
Trinculo: I'm a jester and I work on the King's ship too.

TRACK 2
PART ONE

(In the background we hear the sound of the waves and people talking and shouting. We're at the quayside and the ships are preparing to leave.)

Stephano: That's the King of Naples. *(Alonso steps forward.)*
Trinculo: And that's is the Duke of Milan. *(Antonio steps forward.)*
Alonso: Hello everybody, hello my friend. *(Antonio and Alonso hug each other in a manly sort of way.)*
Prospero: Hem, excuse me…
Stephano: Yes?
Trinculo: Yes?
Prospero: *(Speaking to the audience.)* I'm the Duke of Milan, not him. He is my brother… my very bad brother.
Stephano: Oh? *(As the punctuation marks indicate, one 'Oh' is a question whilst the other is an explanation.)*
Trinculo: Oh!
Antonio: Ha ha ha. Sorry brother! I am strong and you are not. Now I am the Duke and you are a poor man. Now I live in your castle and you live on a small island.

Alonso: Ha ha ha! *(Pantomime 'baddy' laugh. Very exaggerated and theatrical.)*
Antonio: Ha ha ha!
Prospero: Now my bad brother is the Duke and I live on this small island with my daughter Miranda.
Miranda: Yes, twelve years on this small island with my father and Caliban and Ariel.
Stephano: Who is Caliban?
Trinculo: Who is Ariel?
Prospero: They are my slaves.
Miranda: No, they're not, Father. Caliban is a monster and Ariel is a magic spirit.
Prospero: Listen, Miranda, I want to punish my brother and the King. I want to do some magic.
('I want to do some magic' is said almost in a whisper. Very mysterious and with foreboding!)
Miranda: *(Talking to the audience.)* My father has a magic book. He reads his book every day and he does lots of magic.
Prospero: I want to make a storm…a tempest! I want to make a tempest to punish the Duke. *(Rather more hysterical now and excited by his new idea.)*
Stephano: The next day the King and the Duke are in a boat…
Trinculo: …And Prospero makes a magic tempest.

TRACK 3
EXERCISE 4, PAGE 14

Caliban: Ariel, be quiet! I want to listen to the television.

Ariel: Why, what programme is it? Is it football?

Caliban: No, it's the weather forecast. I want to know the weather for tomorrow.

Ariel: Today is a beautiful day.

Caliban: I don't know, it's sunny but it's windy too.

Ariel: Perhaps it will rain tomorrow.

Caliban: Shhh. The weather forecast's starting now.

Weather man: Good evening and welcome to the weather forecast. There will be snow on the mountains tomorrow, but it will be sunny near the sea. It will be cloudy in the afternoon. The next day, stay at home because it will be very stormy: there will be a big tempest!

Ariel: Oh no! A big tempest tomorrow.

Caliban: What's a tempest?

Ariel: Don't you know, Caliban? A tempest is a big big storm.

TRACK 4
PART TWO

(Alonso is standing on deck looking out to sea, the others come out one by one, they say hello and notice the growing storm. In the background we see the butler and the jester cleaning the deck and doing general things. We hear the sound of the waves and the wind, maybe some seagulls too. In this scene the waves and the wind are given voices. Perhaps two or three people, with some interesting microphone effects can make sounds similar to whoosh and crash! If the words they say, or the sounds they make differ from those written, it's not very important.)

Antonio: Good morning, Your Majesty.

Alonso: Ahhh! Good morning, Antonio. It's very windy today.

Wind: Whoosh! Whoosh!

Ferdinand: Good morning, Father, good morning, Antonio. Look at the big waves.

Waves: Crash! Crash!

Gonzalo: There's a big storm today.

Wind and Waves: Whoosh! Whoosh! Crash! Crash!

Everybody: Help!!!

Sebastian: Look! Look!

Wind and Waves: Whoosh! Whoosh! Crash! Crash!

(High up on a hilltop on the island, Prospero is looking down on the shipwreck. Ariel is near him.)

Prospero: Ha ha ha! Very good, Ariel. *(Prospero is more subtle in his laughter than the others but he is cunning all the same.)*

Ariel: Yes Sir, your magic is very good. You are very intelligent. *(Very tongue in cheek.)*

Prospero: Yes, I know, thank you, Ariel.

Ariel: Is the magic finished, Sir? Can I go now?

Prospero: No, Ariel. There is more magic for you to do.

TRACK 5
EXERCISE 1, PAGE 19

Alonso: Good morning, Antonio. Are you sleeping?

Antonio: Oh, good morning, Your Majesty. What time is it?

Alonso: It's half past seven, Antonio. Come on, get up. Today is not a good day.

Antonio: Why? What's the problem?

Alonso: There is a big storm coming. The sky is black and the sea is angry.

Antonio: And where is your son Ferdinand, the prince?

Alonso: Ferdinand is reading a book.

Antonio: And where is Gonzalo?

Alonso: Gonzalo is an old man. He is sleeping.

Antonio: Listen, Your Majesty. What's that noise?

Alonso: It's the storm, Antonio.

Antonio and Alonso: Help!!!

TRACK 6

Listen to the music. Close your eyes and imagine a small boat in the tempest. *(Music)*

TRACK 7
PART THREE

(Ariel goes to the shipwrecked sailors who are lying stranded on the beach of the island.)

Ariel: Everybody is sleeping. I must find Ferdinand and wake him.

(Right from the beginning we hear Ariel's song. First in the distance and then getting closer with the arrival of the spirit. The tune is the same as Frère Jacques and is not necessarily very beautiful as Ferdinand says.)

Ferdinand: Where am I? What's that noise? I hear a song, a beautiful song.

Ariel: Time to wake up, time to wake up.

Ferdinand come with me.

Time to do some magic, time to do some magic.

One two three, one two three.

Ferdinand: It is a beautiful song. Wait for me. Wait for me. *(Ferdinand follows Ariel.. or rather the voice that he hears. He finds himself in front of Miranda.)*

Ariel: Here is Ferdinand, Sir.

Prospero: Very good, Ariel. You can go now. *(Ariel goes away but we see Prospero put a spell on Ferdinand and Miranda which makes them fall in love with each other.)*

Ferdinand: Who are you? What's your name?

Miranda: My name is Miranda.

Ferdinand: You're very beautiful, Miranda. I love you.

Miranda: I love you too.

TRACK 8

See P.26

TRACK 9
PART FOUR

(The four are sleeping on a different part of the island. Antonio and Sebastian wake up. The two baddies in action again. They speak in a very loud 'whisper', still very exaggerated and theatrical.)

Antonio: Ugh! Where are we? Wake up, Sebastian.

Sebastian: We are on an island.

Antonio: Look, the others are sleeping. This is an excellent opportunity for you, Sebastian.

Sebastian: Why?

Antonio: If you kill your brother, the King, you can be the new king of Naples.

Sebastian: Good idea! ...No, wait a minute. Gonzalo is the king's friend. You kill the king and I kill Gonzalo.

Sebastian: OK. Let's do it now, before they wake up.

Antonio: Are you ready?

Ariel: Oh no, this is terrible. I must stop this terrible plan. One, two, three.

Ariel: Time to wake up, time to wake up.

Gonzalo: Where are we? Wake up, Your Majesty.

Alonso: Ugh?

Alonso: What are you doing with your sword, Antonio?

Antonio: Erm… nothing, Your Majesty.

Sebastian: There are lots of dangerous animals on the island.

Antonio: Yes, we want to protect you.

Alonso: Very good, very good. Now, let's find my son Ferdinand.

TRACK 10
PART FIVE

Caliban: Oh poor me! Poor Caliban! Every day I must work for Prospero. I want to be free but I am his slave.

Trinculo: *(voice off stage.)* Where are the others? Where is the King and where is Stephano?

Caliban: Oh no! I hear a voice. It is probably a bad spirit from Prospero. I must hide. *(He hides under his very big cloak.)*

Trinculo: What a beautiful island.

And what is this? It has arms and legs but no head. *(Pause……)* It is a monster! *(A clap of thunder strikes.)* Oh no! Another storm. I must hide. *(He hides under the cloak with Caliban.)*

Stephano: Where are the others? Where is the King and where is Trinculo? And what is this? It has four arms and four legs but no head. It is a monster! Hello monster. Hello!

Caliban: Help! Prospero's bad spirits want to torment me. (Caliban moans and groans when he hears the noise.)

Trinculo: Stephano! Hello! It's me, Trinculo.

Stephano: And who is he?

Caliban: I am poor Caliban. I am Prospero's slave but I want to be free. I want to kill Prospero.

Stephano: That's terrible! You must do something. We can help you.

Trinculo: No, Stephano, it's not a good idea.

Stephano: Yes. We must kill Prospero and take control of the island.

Caliban: Thank you, thank you.

Ariel: I must tell Prospero of this terrible plan.

TRACK 11
PART SIX

Prospero: Ariel, it's time to do some magic. Go to my brother and his friends.

Ariel: Yes, Sir. Of course, Sir.

Gonzalo: Oh, I'm very old. I'm very tired. Can we stop now?

Antonio: Kill him now, Sebastian. He's very tired.

Sebastian: No, not now, tonight when

he sleeps.

(There is strange music and strange shapes who bring in food. They dance around and then leave a banquet for them. With gestures they invite the King to eat and then disappear.)

Alonso: Look at all this food. I'm very hungry.

Gonzalo: Let's eat.

Sebastian: Yes, let's eat.

(There is the sound of thunder and lightning. The banquet disappears and Ariel arrives in disguise.)

Ariel: Stop. You are all bad men. Prospero and Miranda live on this island for twelve years and it's your fault.

(The sound of music and Ariel vanishes.)

Alonso: Oh, yes, it's true. I am a terrible man and now I'm very sorry. My poor brother, poor Miranda.

Prospero: Very good, Ariel.

Ariel: Thank you, Sir.

TRACK 12
EXERCISE 2, PAGE 49

Alonso: It's time to eat but we have no food.

Gonzalo: What can we do? I'm very hungry.

Sebastian: I don't know, what do you want to eat? Alonso, what's your favourite food?

Alonso: Well, I like pasta and pizza. What about you, Gonzalo?

Gonzalo: Oh, I'm an old man, I have problems with my teeth. I only eat soup and ice-cream.

Sebastian/Alonso: Soup and ice-cream! Together?

Gonzalo: No, first the soup and then the ice-cream.

Sebastian: I like ice-cream, too, especially chocolate ice-cream. But I don't like soup.

Alonso: Yes, but Sebastian, what's your favourite food?

Sebastian: I love English food. Fish and chips is my favourite.

Gonzalo: English food is very bad, Sebastian... But we are on an island, why don't we eat fish?

Sebastian: Good idea. If we find some potatoes, we can have fish and chips for dinner!

TRACK 13
PART SEVEN

Prospero: Now, it's time to organise the wedding.

Miranda: Thank you, Father. I love Ferdinand very much.

Ferdinand: And I love Miranda very much.

Prospero: Stop dancing. I must think. Stephano, Trinculo and Caliban want to kill me and my brother is also here on the island. I have a plan, Ariel.

Ariel: Yes, Sir?

Prospero: *(He whispers something inaudible in Ariel's ear.)*

Ariel: Yes, Sir!

(Ariel goes off to the swampy bit where Caliban, Trinculo and Stephano are. He pulls out a piece of shiny cloth and makes them follow it. She sings.)

Ariel: Time to do some magic.

Time to do some magic.
One two three,
Come with me.
Caliban: Look at that. What is it?
Stephano: It's beautiful. I want it.
Trinculo: Me too. Let's go.
Prospero: Now, I will make some magic dogs to scare them. Go! Dogs, go!
Dogs: Woof! Woof! Woof!
Cali/Steph/Trinc: Help!!!
(They all run off with the dogs chasing them.)

TRACK 14
PART EIGHT

Prospero: Now, I must speak to my brother Antonio and the others. Here they are.
Antonio: I don't believe it, my brother Prospero.
Prospero: Antonio, you are a bad man. Sebastian, Alonso you are bad men too.
Alonso: Yes, Prospero, you're right. We are bad men but we're very sorry now.
Antonio: My poor brother, I'm very sorry. Please forgive me.
Prospero: Of course, I forgive you. I forgive you all.
Alonso: I am very happy for this but I am very sad because my son is lost. Perhaps he's dead.
Prospero: No, here is your son. He's not lost, he's here with my daughter Miranda.
Ferdinand: Father!
Alonso: Ferdinand!
(They hug each other etc.)

Ferdinand: This is Miranda. I love her very much. We want to get married.
Alonso: I'm very happy. This is very good news. Let's celebrate.
Prospero: Ah, Caliban, Stephano and Trinculo. I forgive you three, too.
Cali/Steph/Trinc: Thank you, Sir.
Prospero: When I return to Milan with my brother, you can have your island. You can be master.
Caliban: Master of the island? Thank you, Sir.
Ariel: And me? Can I live on the island too?
Prospero: Yes, Ariel. You and Caliban are now free. You are not my slaves. You can do what you want to do.
Ariel/ Caliban: Hurray!
Prospero: Now, everything is organised. Tonight we have a party and tomorrow we must return to Milan together. Ferdinand and Miranda want to get married and Caliban and Ariel can be masters of their island. *(The sound of approval and cheering from all the others who are listening.)*
Prospero: Now I must do no more magic because our story is finished. If you like it, please clap to break the spell.

1. of
2. bad
3. punish
4. storm
5. ship
6. waves
7. happy
8. island
9. and
10. meets
11. make
12. they
13. stops
14. son
15. monster
16. he
17. help
18. food
19. eat
20. bad
21. wedding
22. dogs
23. brother
24. see
25. island

 # Notes

 # Notes

 # Notes

BLACK CAT ENGLISH CLUB
Membership Application Form

BLACK CAT ENGLISH CLUB is for those who love English reading and seek for better English to share and learn with fun together.

Benefits offered:
- *Membership Card*
- *Book discount coupon*
- *English learning e-forum*
- *English learning activities*
- *Black Cat English Reward Scheme*
- *Surprise gift and more...*

Simply fill out the application form below and fax it back to 2565 1113 or send it back to the address at the back.

Join Now! It's FREE exclusively for readers who have purchased *Black Cat English Readers* !

(Please fill out the form with **BLOCK LETTERS**.)

The title of Black Cat English Reader/book set that you have purchased: _____

English Name: _____ (Surname) _____ (Given Name)

Chinese Name: _____

Address:

Tel: _____ Fax: _____

Email: _____

(Login password for e-forum will be sent to this email address.)

Sex: ❑ Male ❑ Female

Education Background: ❑ Primary 1-3 ❑ Primary 4-6 ❑ Junior Secondary Education (F1-3) ❑ Senior Secondary Education (F4-5) ❑ Matriculation ❑ College ❑ University or above

Age: ❑ 6 - 9 ❑ 10 - 12 ❑ 13 - 15 ❑ 16 - 18 ❑ 19 - 24 ❑ 25 - 34 ❑ 35 - 44 ❑ 45 - 54 ❑ 55 or above

Occupation: ❑ Student ❑ Teacher ❑ White Collar ❑ Blue Collar ❑ Professional ❑ Manager ❑ Business Owner ❑ Housewife ❑ Others (please specify: _____)

As a member, what would you like **BLACK CAT ENGLISH CLUB** to offer:
❑ Member gathering/ party ❑ English class with native teacher ❑ English competition
❑ Newsletter ❑ Online sharing ❑ Book fair
❑ Book discount ❑ Others (please specify: _____)

Other suggestions to **BLACK CAT ENGLISH CLUB**: _____

_____

Please sign here: _____ (Date: _____)

Visit us at Quality English Learning Online http://publish.commercialpress.com.hk/qel

Stamp
Here

BLACK CAT ENGLISH CLUB
The Commercial Press (Hong Kong) Ltd.
9/F, Eastern Central Plaza,
3 Yiu Hing Road, Shau Kei Wan,
Hong Kong